Dinosaur Discovery

PTERODACTYLUS

GAIL RADLEY

BLACK RABBIT BOOKS

BOLT

Bolt is published by Black Rabbit Books
P.O. Box 227, Mankato, Minnesota, 56002.
www.blackrabbitbooks.com

Marysa Storm, editor; Catherine Cates, interior designer; Michael Sellner, cover designer; Omay Ayres, photo researcher

Library of Congress Cataloging-in-Publication Data
Names: Radley, Gail, author
Title: Pterodactylus / Gail Radley.
Description: Mankato, Minnesota : Black Rabbit Books, [2021] | Series: Bolt. Dinosaur discovery | Includes bibliographical references and index. | Audience: Ages 8-12. | Audience: Grades 4-6. | Summary: "Learn all about Pterodactylus through diagrams, graphs, powerful illustrations, and fun text"— Provided by publisher.
Identifiers: LCCN 2019034507 (print) | LCCN 2019034508 (ebook) | ISBN 9781623102470 (hardcover) | ISBN 9781644663431 (paperback) | ISBN 9781623103415 (ebook)
Subjects: LCSH: Pterodactyls—Juvenile literature. | Fossils—Juvenile literature.
Classification: LCC QE862.P7 R34 2021 (print) | LCC QE862.P7 (ebook) | DDC 567.918—dc23
LC record available at https://lccn.loc.gov/2019034507
LC ebook record available at https://lccn.loc.gov/2019034508

Printed in the United States.

Image Credits

Alamy: Natural History Museum, 9; Panther Media GmbH, 27; Sabena Jane Blackbird, 16–17; Stocktrek Images, Inc., 12; Dreamstime: Aaron Rutten, Cover; iStock: MR1805, 12, 27; MarkWitton.com: Mark Witton, 19; Science Source: Kurt Miller/Stocktrek Images, Cover; Roger Harris, 28–29; Shutterstock: Carrie Olson, 23; Catmando, 6, 6–7, 8–9, 10–11, 14–15, 20, 20–21, 24, 24–25, 26, 26–27; Elenarts, 19, 24; Herschel Hoffmeyer, 3; Kerry V. McQuaid, 22; matej spiroch, 23; Michael Rosskothen, 6–7, 24, 31; Pyty, 16–17; Ralf Juergen Kraft, 32; Ruslan Harutyuno, 1; Vik Y, Cover; Warpaint, 4–5; twitter.com: Gabriel N. U., 26
Every effort has been made to contact copyright holders for material reproduced in this book. Any omissions will be rectified in subsequent printings if notice is given to the publisher.

CONTENTS

CHAPTER 1

Meet the PTERODACTYLUS

A Pterodactylus sails above the water. Suddenly, its sharp eyes spot a fish. Dinner! Pterodactylus darts down. It dips its long, toothy beak into the water and plucks up the fish. Its sharp teeth pierce the flopping **prey**. The creature flies off to enjoy its meal.

There were many kinds of flying **reptiles**. They're called pterosaurs. Some were small like robins. Others were as big as planes.

What Was It?

Pterodactylus lived at the same time as dinosaurs. It soared above them like a bird. It also laid eggs. But Pterodactylus was not a bird. It wasn't a dinosaur either. Pterodactylus was a flying reptile.

Featherless Flier

Pterodactylus had no feathers. Instead, it might have had a light fur coat. Like bat wings, Pterodactylus wings were made of skin. The wings stretched from its legs to its super-long fourth fingers. The name Pterodactylus means "wing finger."

How Long Was Pterodactylus' Wingspan?

PARTS OF A PTERODACTYLUS

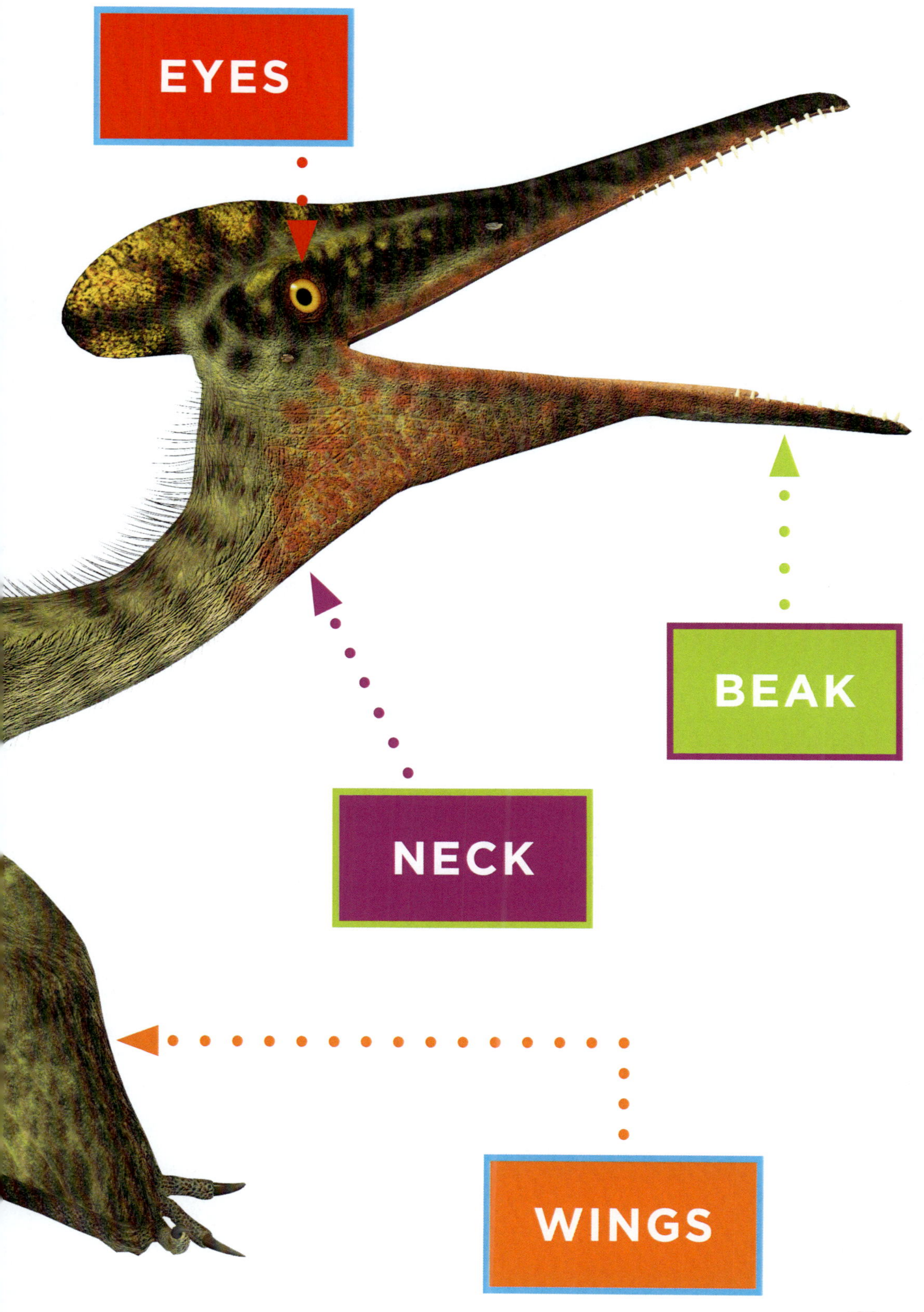
EYES
BEAK
NECK
WINGS

CHAPTER 2

Where and WHEN THEY LIVED

Pterodactylus lived about 150 to 147 million years ago. It ruled the sky during the late Jurassic Age. The oceans were warm at this time. **Ginkgoes** and **ferns** grew along rivers and lakes. Early insects flew with Pterodactylus. Dinosaurs like Stegosaurus and Allosaurus stomped around the ground.

WHEN THEY LIVED

Pterosaurs lived through all three Mesozoic periods. Pterodactylus only lived for a part of that.

PTEROSAURS LIVED
228 to 65 million years ago

TRIASSIC PERIOD	JURASSIC PERIOD
about 251 to 199 million years ago	about 199 to 145 million years ago

MESOZOIC AGE

240 220 200 180 160

MILLION YEARS AGO

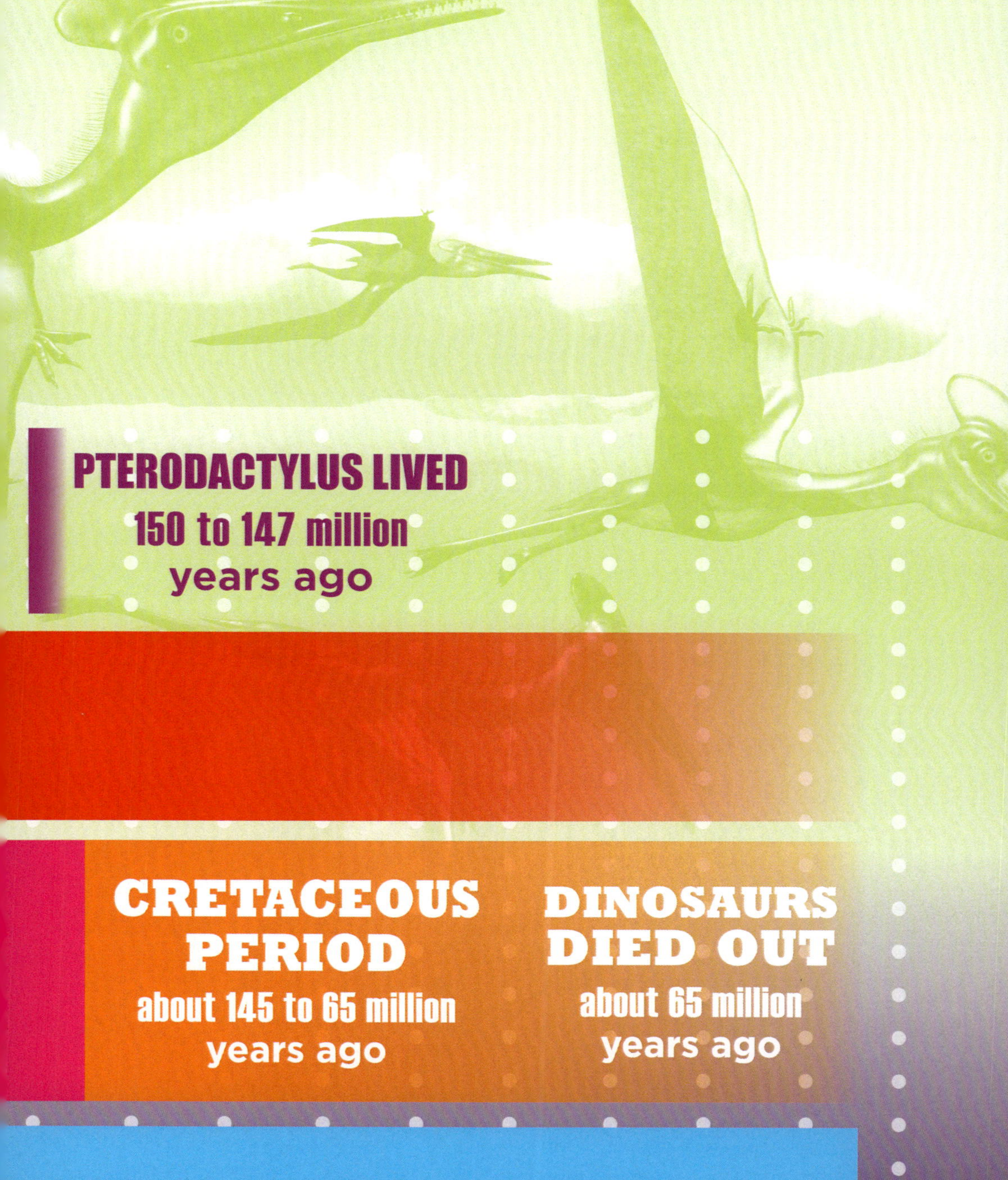
PTERODACTYLUS LIVED
150 to 147 million
years ago
CRETACEOUS
PERIOD
about 145 to 65 million
years ago
DINOSAURS
DIED OUT
about 65 million
years ago
140
120
100
80
60

WHERE PTERODACTYLUS WAS DISCOVERED

Few Fossils

The first Pterodactylus **fossil** was discovered in the late 1700s. It was found in a piece of **limestone** in Germany.

Few Pterodactylus fossils have been found since then. Fossils are rare because Pterodactylus bones were **hollow**. Light bones helped the reptiles fly. But they don't last. They break and mix with dirt.

CHAPTER 3

WHAT THEY ATE

Pterodactylus spent its days looking for food. Scientists believe it usually ate a fish dinner. It would have swooped down to catch fish swimming in the ocean. The flying reptile might have snacked on insects too.

Scientists think larger pterosaurs ate small dinosaurs.

Built to Eat

Pterodactylus' features helped it hunt. The flying reptile's long, pointed beak had about 90 teeth. The teeth could easily stab and hold a fish.

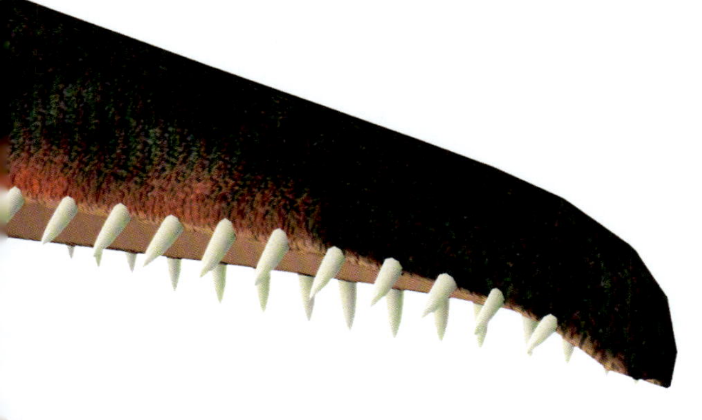

Scientists first thought Pterodactylus' wings were flippers used to swim. Finding more fossils changed their minds.

Like Today's Animals

Pterodactylus shares many features with today's animals. Pterodactylus probably flew like a seagull while looking for food. When hunting fish, it likely dove like a pelican. It probably pecked like a sandpiper when it ate insects.

Animals Pterodactylus Acted Like

TAKING OFF

CHAPTER 4

New DISCOVERIES

Scientists are still learning about Pterodactylus and its family. At first, scientists thought Pterodactylus was clumsy. They thought the reptiles had to leap from high places to fly. Now, their **theories** have changed. They think Pterodactylus leaned forward, rocking onto its folded wings. Then it pushed itself into the air with its arms.

COMPARING PTEROSAUR WINGSPANS

As people search for more Pterodactylus fossils, they also discover new types of pterosaurs. Some of these new flying reptiles are much bigger than Pterodactylus. Others are much smaller.

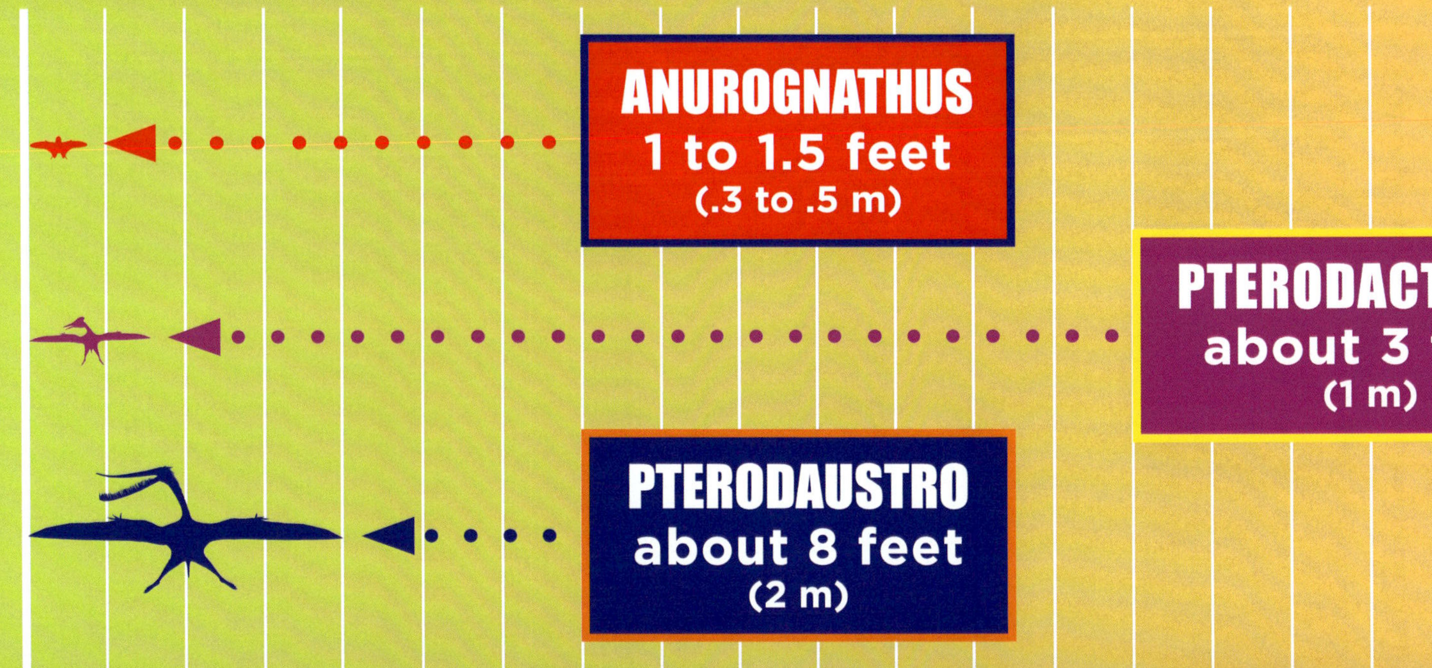

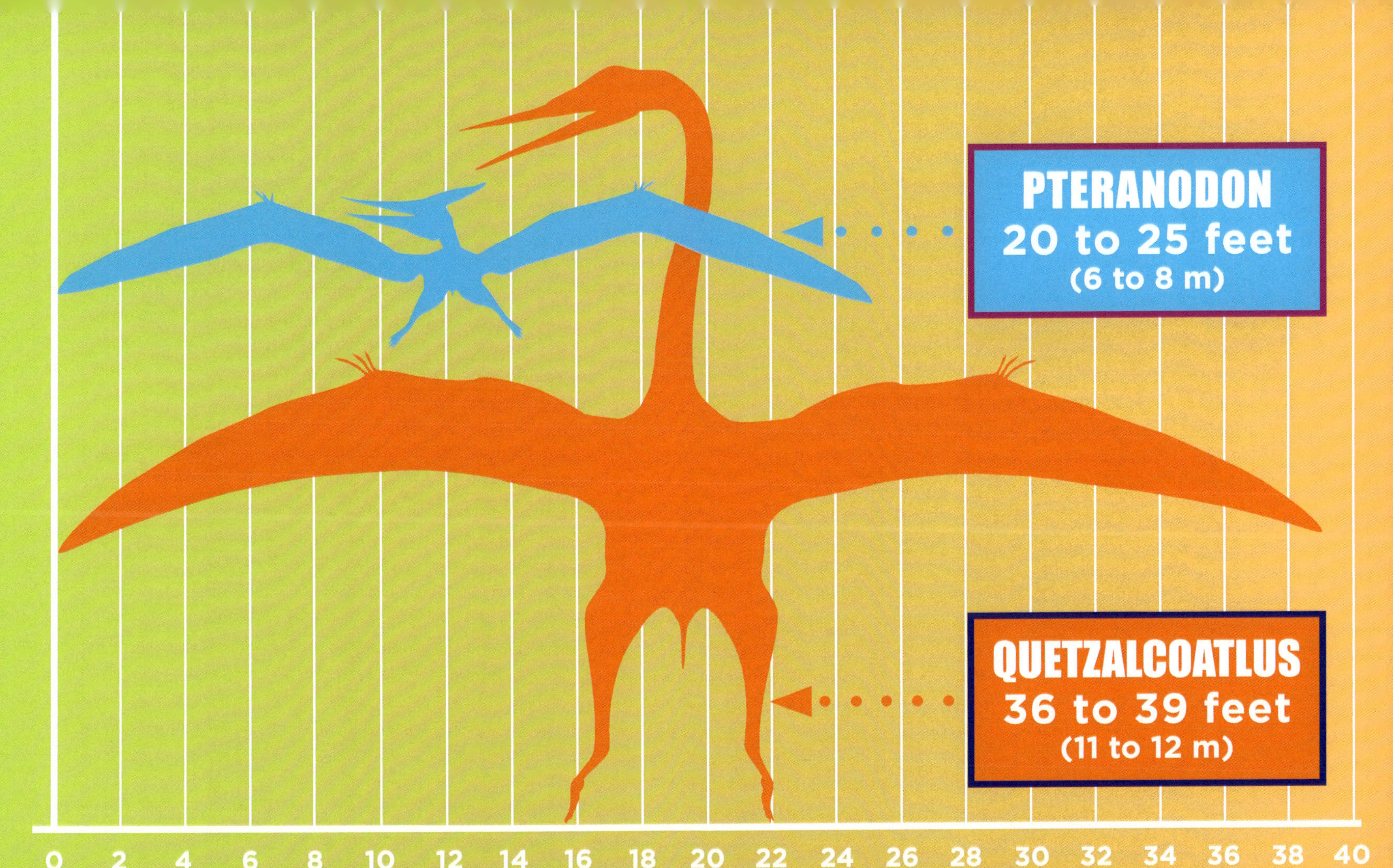
PTERANODON
20 to 25 feet
(6 to 8 m)
QUETZALCOATLUS
36 to 39 feet
(11 to 12 m)
0 2 4 6 8 10 12 14 16 18 20 22 24 26 28 30 32 34 36 38 40
FEET

Always Learning

Studying Pterodactylus isn't easy. Few fossils make it hard to find out more. Scientists keep digging, though. And they use new technology to study old fossils.

Pterodactylus was an amazing creature. Scientists can't wait to unearth more discoveries about it.

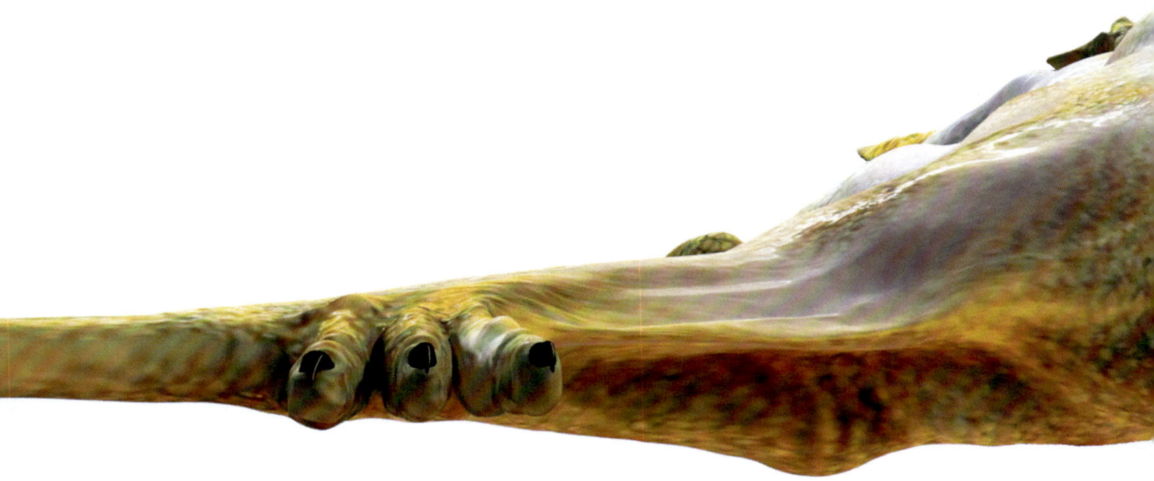

GLOSSARY

fern (FURN)—a type of plant that has large, delicate leaves and no flowers

fossil (FAH-sul)—the remains or traces of plants and animals that are preserved as rock

ginkgo (GING-koh)—a large tree that has fan-shaped leaves

hollow (HAWL-oh)—having nothing inside

limestone (LAHYM-stohn)—a type of white stone that is commonly used in building

prey (PRAY)—an animal hunted or killed for food

reptile (REP-tile)—a cold-blooded animal that breathes air and has a backbone; most reptiles lay eggs and have scaly skin.

theory (THEER-ee)—an idea or opinion that is presented as true

BOOKS

Carr, Aaron. *Pterodactyl.* Mighty Dinosaurs. New York: AV2 by Weigl, 2019.

Rober, Harold T. *Pterodactyl.* Dinosaurs and Prehistoric Beasts. Minneapolis: Lerner Publications, 2017.

Weakland, Mark. *Flying Reptiles: Ranking Their Speed, Strength, and Smarts.* Dinosaurs by Design. Mankato, MN: Black Rabbit Books, 2020.

WEBSITES

Fun Dinosaur Facts for Kids
www.sciencekids.co.nz/sciencefacts/dinosaurs.html

Prehistoric Animals
kids.nationalgeographic.com/animals/prehistoric-animals/

Pterodactyl – Kids
kids.britannica.com/kids/article/pterodactyl/390080

INDEX